Bugs

Centipedes

by Trudy Becker

FOCUS READERS
PIONEER

www.focusreaders.com

Focus Readers is distributed by North Star Editions:
sales@northstareditions.com | 888-417-0195

Produced for Focus Readers by Red Line Editorial.

Photographs ©: Shutterstock Images, cover, 1, 4, 6, 8, 10, 12, 14, 17, 18, 20

Library of Congress Cataloging-in-Publication Data
Names: Becker, Trudy, author.
Title: Centipedes / by Trudy Becker.
Description: Lake Elmo, MN : Focus Readers, [2023] | Series: Bugs | Includes index. | Audience: Grades 2-3
Identifiers: LCCN 2022028386 (print) | LCCN 2022028387 (ebook) | ISBN 9781637394489 (hardcover) | ISBN 9781637394854 (paperback) | ISBN 9781637395585 (pdf) | ISBN 9781637395226 (ebook)
Subjects: LCSH: Centipedes--Juvenile literature.
Classification: LCC QL449.5 .B43 2023 (print) | LCC QL449.5 (ebook) | DDC 595.6/2--dc23/eng/20220622
LC record available at https://lccn.loc.gov/2022028386
LC ebook record available at https://lccn.loc.gov/2022028387

Printed in the United States of America
Mankato, MN
012023

About the Author

Trudy Becker lives in Minneapolis, Minnesota. She likes exploring new places and loves anything involving books.

Table of Contents

CHAPTER 1

A Place to Hide 5

CHAPTER 2

Legs All Over 9

CHAPTER 3

Centipede Skills 13

THAT'S AMAZING!

New Legs 16

CHAPTER 4

Centipede Life 19

Focus on Centipedes • 22

Glossary • 23

To Learn More • 24

Index • 24

Chapter 1

A Place to Hide

A centipede speeds across the dirt. Its many legs move quickly. The centipede finds a rock. It slips under. It stays there until nighttime.

Centipedes live in dark, damp places. They can live all over the world. They hide under rocks, logs, and soil. Then they come out to catch **prey**.

Fun Fact

Some centipedes live in deserts. Some live in forests. Others live in people's houses.

Chapter 2

Legs All Over

Centipedes have flat bodies and many legs. These legs help centipedes run fast. Each pair of legs connects to a body **segment**.

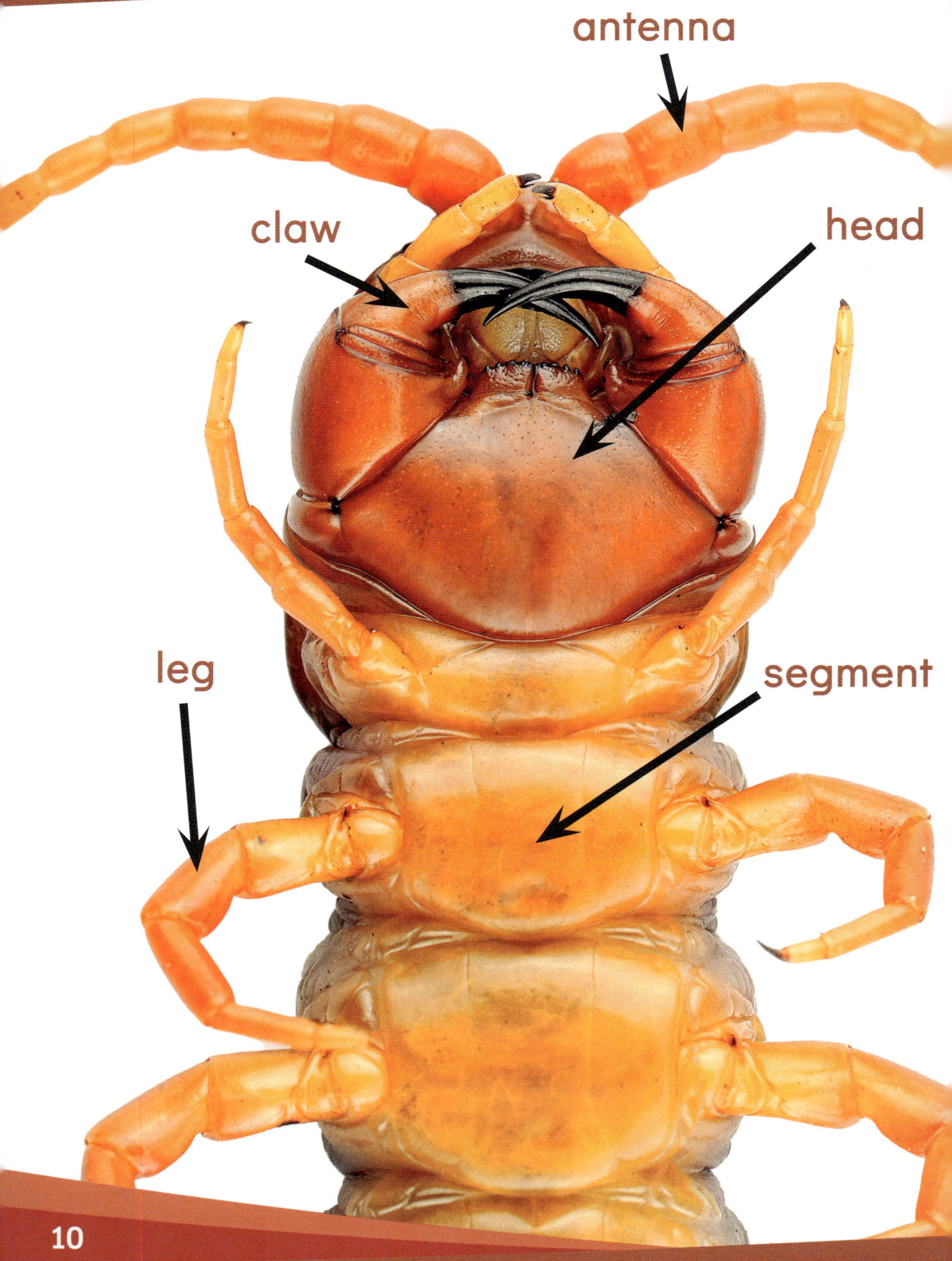
antenna
claw
head
leg
segment

A centipede has two long **antennae**. It has jaws for biting and holding food. It also has claws. They help the centipede catch and kill prey.

Fun Fact

Some centipedes have 15 pairs of legs. Others can have up to 191 pairs.

Chapter 3

Centipede Skills

Centipedes move around at night. Their eyes don't see well. But centipedes use their antennae to smell and touch. That helps them hunt.

Centipedes use **venom** to kill their prey. First, centipedes catch their food. Next, they use special claws to put venom into the prey. Then the prey stops moving.

Fun Fact

Centipede venom doesn't work right away. So, centipedes wrap their bodies around prey while waiting.

New Legs

Sometimes centipedes lose their legs. Their legs could get stuck. Or a **predator** could get them. But centipedes can fix that. When they **molt**, they grow new outer coverings. They can grow new legs then, too.

Chapter 4

Centipede Life

Centipedes are **carnivores**. They usually eat insects, spiders, and worms. But centipedes are prey for larger animals. Birds, toads, and mice eat them.

In spring or summer, centipedes lay eggs. The mothers protect the eggs. When the eggs hatch, the babies have only a few legs. Then they molt and grow more legs. Soon, they can start to find food.

Fun Fact

A centipede can live for up to six years.

FOCUS ON

Centipedes

Write your answers on a separate piece of paper.

1. Write a sentence that explains the main idea of Chapter 3.
2. What would you do if you saw a centipede in the wild? Why?
3. How do centipedes kill their prey?
   A. They use venom on their prey.
   B. They crush prey with their bodies.
   C. They catch prey in a web.
4. Why don't centipedes eat plants?
   A. They live in areas with no plants.
   B. They are carnivores.
   C. Plants are too dangerous.

Answer key on page 24.

Glossary

antennae
Long, thin body parts on an insect's head. The parts are used for sensing.

carnivores
Animals that eat meat.

molt
To lose an older covering so a new one can grow in its place.

predator
An animal that hunts other animals for food.

prey
Animals that are eaten by other animals.

segment
A part of an animal's body. Some animals have bodies with many similar segments.

venom
Poison that comes from an animal's sting or bite.

To Learn More

BOOKS

Hansen, Grace. *Giant Centipedes*. Minneapolis: Abdo Publishing, 2021.

Nickel, Scott. *World's Grossest Creepy Crawlers*. Minneapolis: Lerner Publications, 2022.

NOTE TO EDUCATORS

Visit **www.focusreaders.com** to find lesson plans, activities, links, and other resources related to this title.

Index

A
antennae, 11, 13

E
eggs, 21

L
legs, 5, 9, 11, 16, 21

P
prey, 7, 11, 15, 19

Answer Key: 1. Answers will vary; **2.** Answers will vary; **3.** A; **4.** B